Acknowledgements

The greatest adventure of all Emma, Natalie and Wil

My good mate and fellow adventurer Chris

Front Cover
Exploring the Catalan Pyrenees

ISBN: 978-1-312-66385-5

Copyright: © 2021 Standard Copyright License
Language: English
Country: United Kingdom
Version: 0.1

Preface

What is an adventure?

I have a low boredom threshold, shockingly low, but I am also able to become disproportionately enthused by a challenge, so excited in-fact that just doing something I have never done before becomes an adventure. Succeeding is a win, my mental state 'leapfrogs' any and all negativity and I am buoyed until the next great challenge. Adventures are my way of coping with a world that can sometimes bear down on me, such that, if left unchecked, could easily crush me.

I cope by having adventures, every day, cope is the wrong word, it implies that I scrape by. Not so, my adventures are brilliant, awesome, exciting and new.

They are not a way of surviving; they are my way of thriving.

Contents

1	You don't have to be Indiana	4
2	What is worthy…..?	10
3	Essential Elements	24
4	Mix the physical and the mildly cerebral	29
5	When you feel a bit more confident	32
6	What does this all mean?	49

Adventure

1

You don't have to be Indiana

Before we start, it is vitally important that we have a shared understanding of exactly what I mean by adventure. We could, wrongly in my view, be led to believe that adventures are high adrenalin, dangerous feats of endurance and 'daring do'. Bear Grylls climbing the Matterhorn drinking water from thawed snow (the yellower the better), Steve Irwin catching snakes, Indiana Jones battling giant insects, magic, villains and the countless other 'real' and fictional characters who grace our screens with explanations of how they do every day what we never will. These are not adventures, these are theatre. Fabulous, entertaining and often really compelling theatre, but theatre, nonetheless. This is not the kind of adventure that I am going to talk

Adventure

about, instead I will focus on what you can do, yourself, alone or with close friends, without the help of a film crew, producers, location caterers or on-site private physicians.

So, what is an adventure, in my parlance?

It starts with that seed thought, one of these little questions.

"I wonder what it would be like to"?
"I have never been to"?
"How would I feel if"?

You then return to your 'child mind', it is still there, buried deep, but it is still there.

Let me give you some examples, to whet the appetite.

"I have never been to (input a local attraction here), I wonder what it is like"?

I recently noted that I have never walked up Glastonbury Tor and looked over the Somerset Levels. I remarked to myself that this was likely a view that Sir Lancelot would have savoured. And it is only 8 miles from where I live.

I planned a half day
I read a little about the place
I double checked where I could park my car

Adventure

On the morning I planned to go,

I checked the weather, put a waterproof in my backpack
Packed a drink
Made one sandwich

I left at 10:30 (having done my VAT return)
Drove and parked, where I had planned to park
Walked to the foot of the hill
My march to the top periodically slowed by my hitherto lazy aching legs

By 11:45 I was sitting at the top, panting.
The view was splendid
After about 15 minutes, sat on the grass, I ate my sandwich and enjoyed my drink

Sitting atop this wonderful magical mound I pondered this place, steeped in folklore and history, why was it such a magnet for Christians and Pagans alike. Was it really visited by Jesus or did Joseph of Arimathea come alone (more likely) in his search for Tin? How severe was the Earthquake that destroyed the 14th century church (St Michael's)? And before this, it was a lookout post for a Roman garrison, they must have hated walking up here every morning. Just how grizzly must the scene have been when, during the Reformation the Tor was the site of the execution (by hanging) before the last Abbot of Glastonbury was Drawn and Quartered, with two of his fellow monks? Was this where Lancelot came with Gwynevere (Arthur

Adventure

wasn't too pleased about this), and is there really a cave beneath the Tor, through which one can pass into the realm of Fairies? All of these thoughts cantered through my mind as I sat looking over the Vale of Avalon. At 13:30 I started to walk back down the hill. By 15:00 I was home, back at my desk, doing another tax return. On this day I had had an adventure, I had wondered what it would be like to do something new, I had researched a little, I had planned a little and I had done it, I had achieved something.

This was an adventure, no risk to life, nothing worthy of a dramatic backing track, narration or radical camera angle but I was richer for the experience.

This book is about adventures, everyday adventures that are unique. We can all be adventurers; this book is as much about frame of mind and 'Spirit'. Everyday can be an adventure, if you want it to be. The only thing I can guarantee you is that freeing some of that 'Spirit of Adventure' will liberate endorphins to course through your veins, you will soon become addicted.

So, in the simplest terms an adventure is something which challenges you to do something different. Slightly different is enough.

If you don't go out a lot, a well-planned walk around the block is a massive adventure

Adventure

If being in a crowded place is worrying to you then a well-planned trip into town (with an escape plan) is massive.

A night out at the theatre is an enormous adventure, especially if it is new.

Painting a picture, writing a poem, baking a cake.

All of these things are adventures, all of these things leave you beaming, endorphins flowing and success coursing through your veins.

Yes, yes, yes, climbing the Matterhorn, sailing single handed around the globe are also adventures, but no more worthy than you or I doing something new. What for someone else is 'normal and everyday' will not be for you and me. What excites you, they may not understand, why should they, it is not their adventure, it's yours.

Here are some examples of adventures. I dare you to try one:

- Find a local attraction and visit it
- Explore some local history
- Do something that scares you a little
- Do something you do every day, differently
- Don't drive, get the train, bus or cycle (if you have access to a bicycle)
- Ask yourself questions and then explore the answers

Adventure

- Do a little research and then 'Go and See'

As you get more comfortable with your successes, escalate… Start with a few hours. Then a day, then two, then more.

Adventure

2

What is worthy…..?

There is no activity which is not worthy of the term adventure. It is totally in your gift to define what, for you, is an adventure. It is all about how you approach it. I have already said that the majority of what we are shown on the TV is NOT adventure it is theatre.

Avoiding or ignoring the temptation to make your adventure a major feat of physical or mental endurance, instead setting yourself goals which push your boundaries, just a little, is a sure way of building confidence as an adventurer. Remember this is for you.

There is no exhaustive list but here are a few of the things which I consider to be adventures and have approached enthusiastically.

> **Going on walks** that start from home, doing something without the use of a transportation

Adventure

system other than that which you are blessed, whether legs, prosthetics, wheels, wings, jet packs, transcendental meditation. With the help of a local map, find out what is on your doorstep. If you have an Ordinance Survey Map or can access maps online (or with a phone), have a look at what the local Footpaths and Bridleways have to offer. Plan yourself a walk that takes you on a circular journey of discovery in your area. If you live in a city or town do the same, plan a circular walk. Plan in a reward break, stop, eat a snack, have a drink and carry on. You will, I promise, be amazed at just what there is within a mile (1.6 km) of your home. If accessibility is a challenge additional route planning may be required but there will be journeys to be had. On returning from your trip, you WILL feel enriched and more aware of your home. This will be an adventure of discovery. It will whet your appetite if only to do another. My wife, Emma in I had lived in our village for 9 years before we ever properly started to discover it. We got the map out and found the area riddled with Footpaths and Bridleways. First, we planned a walk to a pub in the next village (The George), this was a first, this was an Adventure. A little more planning done, and another walk was arranged, from our home to our home, this time it was to be most of a day. The map showed us may interlinked footpaths. We checked the weather, packed snacks and water and set the goal of getting to a certain point before stopping for a snack (a

Adventure

little bridge over a stream). The walk again showed us the area we live in from a completely different angle. It felt, to us, like a discovery, each corner brought a smile, each vista a sharp intake of breath. We saw and talked about the plants and wildlife that lives on our doorstep, we sat on our bridge eating a sandwich, like contented children, we shared a drink of water, happy at having achieved the halfway goal. The return took us up a hill under the leafy canopy of old oak trees, back to the Footpath that would take us home. What an achievement, we had walked for hours, full circle, we had found an old 'Stately Home' we never knew existed, looked over the Somerset Levels from The Isle of Wedmore, eaten home-made food and made it home in one piece. This was another Adventure, something we had never done before, something we had discovered.

Big walks that you have travel to, whether a trail or to see with your own eyes, a little piece of history (my trip the Glastonbury Tor). Wherever you live, in the country or a town, there will be adventures to be had, and there are many websites and books which will tell you where you can go exploring. The key is to set yourself a goal, a route, and have a plan, this way you will have something to achieve and something to celebrate. Ensure you factor in the time it takes you to get there and back. They require just a tiny bit more planning. But if you get the planning right the feeling of elation you get form having it all come together smoothly is awesome. These composite parts are each a goal to be savoured. An example of one of our all-time brilliant Adventures was when we lived in Plymouth. A 5-minute walk from our flat to the Cremyll Ferry (a foot ferry from Stonehouse in Plymouth to the Mount Edgcumbe estate across the river Tamar). From this amazing stately home, we followed the footpath to Cawsand where we rewarded ourselves with Ice Cream before then catching the bus back to the Mount Edgcumbe estate and the last ferry back. We discovered an historic ferry ride, a beautiful estate home, a marvellous clifftop walk and had ice cream in a stunning fishing village, the bus ride home was filled with top deck views, the final ferry ride home ended with a pint in the Grape Vine pub where we all congratulated ourselves on our achievement. 20 years on the Children still talk about it as 'a great Adventure' (we carried them most of the way).

Visit a monument or attraction but do the reading first, this makes all the difference, a tiny bit of research invests you in the adventure! Just finding out a little bit about a place before you go and find it, makes the discovery so much better, you understand it more, you contextualise it better. I remember having a few hours to kill in London, I had been 'head down' solving client problems for weeks and I

Adventure

needed a distraction, I was staying in The City. Out came the AtoZ map, I found the

Monument to the Great Fire of London (1666). I read a little about the monument, some of the things that had happened there over the years and looked at photographs of the stunning sculpture on one of the pedestal panels.

In the morning, after a client meeting, I set off in search of the junction between Monument Street and Fish Street, to see the sculpture was my goal, to actually climb the monument was to be a bonus. I achieved both, I was back in good time for a client meeting in the afternoon, my mind was fresh, enlightened and I had achieved something out of the ordinary (for me at least), I had researched, set off on a voyage of discovery and I had achieved my goals. This was an Adventure.

A cycle trip, the more walking adventures you do, the fitter you will become, or maybe you are fit already. Cycle adventures are 'completely different' because you are using a new mode of transport, but you are still the 'engine', it is still all your work. Set a route, set a destination and return, stop for lunch, take your time.

That bicycle you own, it has been there for years, untouched. Nagging you, or at least making you feel a little guilty each time you

Adventure

had to move it to get to the boxes stored behind it. But be careful, we always underestimate how many miles our backside, legs, back and lungs can manage. Plan a circular route or a clear halfway point from which to turn back. Remember, people wear padded shorts for a reason and turn back long before you are too tired to continue. Cycling is awesome, but more tiring than it looks. Like walks, start small and build up. Also be safe, helmets, bright clothes, lights and working breaks, bells are all essential. Big roads and narrow 'A' roads can be really frightening and very dangerous so plan your journeys for safety and keep the distances low. But when it works a cycle trip is super rewarding, if you have planned the way you plan your walks you will have the added bonus of having travelled a little further and faster, under your own steam.

Be aware that you are sharing the roads with others, cyclists, sadly due to a minority, have earned an unpleasant reputation of being selfish road users, do not become that person, there is plenty of room for everyone, and your adventure will be much more rewarding if not tainted by argument. If you have a way of carrying things on your bike fabulous, small backpacks are great but do not underestimate how hot you can get cycling, pack light. This rule is good to observe whatever your chosen adventure. But for Walking and Cycling it is essential. Timing is also the key, I once set

Adventure

myself the challenge of cycling from home to Glastonbury 8 miles, I had never done it before, so I planned my route and set off. In Glastonbury, I locked up my bike and went for a stroll, serendipitously I bumped into a friend, the lovely Dymphna, so we went for a coffee. After a couple of hours chatting, we said goodbye and I jumped back on my bike to cycle home. It was Autumn, the nights were drawing in. I found myself cycling home across the Somerset Levels with no lights on a dark night, along small dark lanes. I genuinely feared for my life. Planning is everything.

A trip in your car, to somewhere you have never been, to watch a sunset. We are all very well used to travelling great distances in our cars or using our cars to go to and from work. But they can also bring us so much more, and in great comfort. So that mini adventure to a place, town, market, monument, museum easily becomes and adventure when you plan it as an adventure. Again, plan it from home. Choose a goal, take your time, set your reward stops. In my youth I would ridicule the people who would "go for a drive" get there, have their picnic and drive home. My own grandparents would drive from Tiverton to Budleigh Salterton, park overlooking the sea, eat a sandwich and then drive home. But surely this was a journey? Ah but when it was to somewhere else, somewhere different, it was a planned adventure, they got home having achieved

Adventure

something. Emma and I have regularly researched great gardens to visit, learned something of their history and then made the journey. Even if only a 45-minute drive, these days can be wonderful adventures.

- Westonburt Aboretum
- Kilver Court
- The Quantock Hills

The list of driveable adventures within an hour of where YOU live, will be enormous, the key is that you plan something which is not the ordinary 'go to work' trip. Not only will you achieve a mini adventure, but you will also appreciate your car more. A car used for a pleasurable, different journey is brilliant. 'ROADTRIP'!!!! And if you have to make a long Road trip for work or otherwise, plan the journey as though it were an adventure. Route, Food, Stops. Look at a map. Take note, from a map or online, of some of the points of interest en-route, this will make an ordinarily boring journey, interesting. I love a long drive, because every long drive, with a little bit of planning as fascinating. I bought a little car fridge from Halfords (other shops available), this means that I can have fresh food and a cool drink at any time. But I also plan breaks and goals. And NEVER drive for too long, it makes the adventure tiring, it is unsafe, and it spoils the journey by making it a chore. Plan for stops, plan to stop for the loo!

Adventure

Travel somewhere by bus or train, look out of the window, when was the last time you used public transport to go somewhere for pleasure? Armed with a timetable, transportation systems are enormously liberating. A train journey or a bus journey, to a new place, look out of the window, drink in the views without the worry, stress or 'work' of driving. Look up from your phone and out of the window. Get to where you are going, enjoy being there and then use that same transport to take you home. If changes are required, plan them, it adds to the adventure. The amazing thing about these adventures is that you will be surprised at how a little bit of planning rewards you with a trouble-free adventure with minimal stress or energy expenditure. Near our home there is a Bus Stop, but it only gets 1 bus every two hours. We started by seeing where the routes went from this stop. From here we planned a day out in the city of Wells. We got the bus, looking out of the window we passed through Hamlets and passed houses we had never remarked before. We visited the City of Wells, had a wonderful lunch at the The Crown (made famous in the movie Hot Fuzz), an afternoon drink at The Swan and got the bus home, an adventure from the doorstep.

Go shopping, to a place you have never shopped, especially if public or crowded places make you a little nervous. I find that I can get more than a little distressed in crowds, especially if they are 'milling around, Christmas markets are the worst. If well planned I can go and return and feel enormously proud of my success, often more so than many other adventures, but I am not resilient enough to do these often. Be careful. These for me can be more frightening than some great feats of 'daring-do'. Don't let anyone suggest to you that what they find easy, isn't a major feat for you. For an even mildly agoraphobic person (where this agoraphobia manifests itself as a fear of crowds) such a journey is enormous. You may want to plan an escape route (I always do) and I take a companion with me, without Emma or one of my children I simply would not be able to take on such an adventure, especially if there is a risk of the place becoming busier than anticipated. PLAN.

Go to the theatre, a movie, a concert, a festival, like the Go Shopping adventure, be aware of how the environment might make you feel, for the vast majority these will seem like straightforward trips. But, if you are like me, plan your escape routes, if only in your mind. If you have done this planning, your escape plans will be reassuring. A couple of summers ago Emma and I planned to go the Bristol Downs Festival, we planned the entire day, a drive, a Bus ride, the Festival, in time to see Seasick Steve play, and then a rapid escape. The day was brilliant, beautifully timed,

Adventure

challenging but fantastic.

So, the key to successful adventures on your doorstep is to: **Plan, go somewhere, learn and or experience something, come home richer**

But there are other adventures!

Read a book, I am dyslexic, many many people are, for dyslexics, reading a book is tremendously difficult. If this is you, find something 'light' to read, put aside time, make your cup of tea, sit down and read it, cover to cover. If you need many breaks, take them, this is not a competition. In the UK (at least for my generation) the system of education was specifically designed to 'make you feel stupid' if you found reading hard work. Start with a Graphic-Novel (or a comic), as an adult, there is nothing wrong, and indeed everything right, with reading a children's book (many a fool, calls 'The Little Prince' a children's book). Better to start with Sue Townsend and enjoy it than Fyodor Dostoevsky (or Dr Yeskey, as my sister requested from the University of East Anglia library on her second day after receiving her "reading list") and hate it (when you get more practiced at reading you will love Dostoevsky, if that day never comes, it's OK). If you are new to reading books, the feeling of elation and success when you finish, will be as

Adventure

great if not greater than a hike up Mount Everest, and what you read will be chiselled into your memory permanently.

If you are a monstrously busy person (like my wife Emma) plan a day at home, in front of the fire, with a book and escape to another world. Adventure from your armchair.

Write, a book, a poem, a diary, a blog, just write. The best writers I know, a year before they wrote anything, would have told you they couldn't (many of them are also Dyslexic).

Are you shy? Try and join a group, sport, study, shared interest, take it gently, tell the organisers that you are shy, take it one step at a time. The organisers will be supportive, they will encourage you. Joining a group if you are shy is an enormous and onerous exercise, but once done you will feel awesome. Plan it well and it is an adventure, you can do this to Learn a new skill – contact the local college, plan to do a course which will enrich you.

Cook!!! It is exciting, rewarding and adventurous. Bake something, when was the last time, if ever, you took raw materials and transformed them into something? What is the worst that could happen? In fact, any form of transformative activity is both rewarding and highly creative, be creative, plan it and push your boundaries, this is adventure. Don't

Adventure

believe me? Try it……

For the more traditional amongst you:

Plan a **camping trip**, the joy of sleeping out in the open and waking up closer to nature and if you have never done it before, it's just fabulous. But plan it, food, warmth, TOILET PAPER,

Toilets or a spade. First Aid Kit etc. I treat the planning and preparation as part of the adventure. So, getting ready is fun.

Go canoeing for a few hours, or a day (*Travel a thousand miles by train and you are a brute; pedal five hundred on a bicycle and you remain basically a bourgeois; paddle a hundred in a **canoe** and you are already a child of nature. ... **Pierre** Elliott **Trudeau***).

My mate Gordon having a paddle on the Wye 2016

Adventure

There are many leisure companies who run canoeing centres on rivers across Europe. Find one, contact them and plan a Canoe trip. I am so hooked on these that my last adventure was a 19-day 500km solo adventure (Reflections on Water, available from Amazon), I am not suggesting you do this, an hour's excursion can be as magical as traversing the great lakes.

If you have a **motorcycle**, ride it somewhere, that is not work. Ride it slowly, open your visor, smell the air. Riding or driving to 'get to a place' is a chore, make the journey part of the adventure. After all, this is why you bought it, isn't it? My friend Chris and I circumnavigated the four compass points of the UK mainland in 2006 (The Liquorice Road, available form Amazon), it was a spectacular journey, we discovered parts of this country we had never considered, enjoyed vistas and swept through landscapes worthy of Tolkien. Another time we journeyed across France over the mountains to Pamplona and the Fiesta de San Fermin (The Bull Run). On another adventure we took our dirt bikes Off-Road over the Catalan Pyrenees (we actually tried twice but the first time my motorbike died up the mountain, even this proved to be a memorable adventure). These were truly epic adventures, but superbly well planned and hence great fun. As big a feat for me as braving the Bath Christmas market.

Essential Elements

For an adventure to truly be successful it must not be a coincidental or haphazard event. Planning is everything. In fact, good planning builds the anticipation, it fuels the reward and reinforces the feeling of success when things go to plan and when things go wrong, and your disaster planning pays off, even better.

The level of planning will necessarily be commensurate with the adventure being envisaged. If like me, you become a 'mini-adventure' nut then you will enjoy the planning as much as you do the actual Adventure.

Your planning will entail:

- Internet and bookish research
- Maps and satellite surveillance
- Kit lists
- Numerous attempts and practice on how bags will be packed

Adventure

- Training, acquiring the skills to keep you safe
- Trial runs

Now obviously if you are planning a day trip, in the car to a sea view on the North Devon coast, you won't be doing all of the above, at least not to the 'nth degree' but it is worth having some of the above as a default list in your head. I have all of the above as the minimum.

In fact, my basic list first includes these 6 items, and then I consider the above:

- How long am I going for?
- How will I get there?
- Where will I stay?
- What will I see, experience, learn, when I am there?
- Do I have the skills?
- What equipment will I need?
- How will I stay comfortable?

If you think this is all very obvious then consider the chap who left his home near Bournemouth for a canoeing Adventure on the River Spey in Scotland. A seasoned canoeist, a man blessed with years and years of canoeing experience, an ex-military expert, in everything, an Ex-Marine, who, as well as being the world's most arrogant IT director and a complete bully, knew everything. A man who did not need a checklist or a plan, for this gentleman such a journey was to be second nature. The rest of the group who

Adventure

were to participate, despite never having met each other face-to-face had agreed to meet up the day before setting off. I found out that he had signed up completely independently from me, can you imagine my disappointment when he appeared, at the Cairngorm Hotel in Aviemore. In the months and weeks before the journey we had meticulously planned, exchanging calls and emails, regularly discussing.

- Exactly what kit we needed, repeatedly comparing notes, "ahh great you have a spare torch, I have a spare Kelly Kettle" etc.
- What food we should we carry
- Accommodation (Tents, Hammocks, Tarpaulins)
- Fire Starting, who's got what?
- First Aid
- Spare Paddles
- Truly to the 'nth degree'

At each stage we had shared the list, asked the questions and invited input. His response had always been 'Pfffff, I have it sorted, don't worry about me, I was a Royal Marine.

As we all sat in the bar of the Cairngorm, looking out at the car park, Julie asked, "Peter, is that your truck just there"? "yes" came the proud reply. "I just bought it last week, bargain, blah blah blah". Julie pondered the response for a moment or two and then popped the $64,000 question. "where is your canoe"?

Adventure

If I can answer the planning questions (my key planning questions) I can plan for most adventures and occurrences. I have, however, already said that an Adventure is what YOU make it and that you should start small. If your Adventure is a trip to the theatre or a country walk or a drive, then canter through this list and discount what is not relevant, as your Adventures become more challenging, then planning, by default, becomes more exacting. When I go walking for a day, I plan to travel light, carry water and a snack. 19 days living out of a canoe and I need all of the above and more! A day trip to Dawlish, I need a bus timetable, a train timetable, tickets and enough money for a Pie and a Pint when I am there. A two-week Pyrenean adventure on Dirt Bikes requires, transport, trailer, bikes, fuel, tools, replacement parts, tents, sleeping bags, food, camera, toilet rolls, spade (for use with the toilet roll), drink, clothing and the list rightly goes on. But the planning is enormous fun. Testing how you will pack bags is both interesting and vital. As is planning routes and stops.

The essential in all of this, however, is to plan just enough to allow you to BE A KID AGAIN.

When I was a small boy, I was always quite independent, up to the age of 11, we lived in Tiverton in Devon, in a great big house called Cowley Lodge on Blundells Road. The house backed onto some fields, a Blue Bridge, over the River Lowman and the end of an old railway track, across the road was another old railway track and, strangely, above this, the Canal. The place was made for adventures, every Saturday morning I would make and pack my

Adventure

Marmite Sandwiches, fill a bottle with Orange Squash and set off to cross the Blue Bridge and walk up the banks of the River Lowman until I got to the forest (it was actually a small wood, but I wasn't very big, so to me it was Fangorn). There, other adventurers would come, we would meet friends, build Dens, Dams and Fires. We would laugh, cry, climb trees, fish for Lampreys and Bullheads. It was the world of Swallows and Amazons; it was another world, where Adventures were boundless. I think it was here that I learned to relish everything I did.

My daily walk to school wasn't so different, out of the back of the house, over the Blue Bridge, across the field and up the old disused railway track, until I got to the Elmore County Junior school, about a two-mile walk. "A two-mile walk!!!??? Up an old Railway track!!!!???? At the age of 8 - 11! You wouldn't let a university student do that nowadays. My nickname was Squelch, because I used to play in the river on the way to school, squelch was the sound my shoes made by the time I got to there. In the summer I could catch Slow Worms on my way home, in fact the walk home could take hours and hours.

As a result of this beginning, I seek out the experience in everything. I urge you to do the same. It makes everything exciting; it makes everything fun; it makes everything an Adventure.

BE A KID AGAIN

Adventure

4

Mix the physical and the mildly cerebral

I have, a few times, mentioned how important it is to do a little research before setting off. The logic for this is simple and disproportionately rewarding. Just a few minutes of reading or research will transform your experience of a place.

When visiting the small village of Wedmore in Somerset, you will remark to yourself, "what a pretty little village" but with a tiny bit of research you would understand that when Alfred (a King of Wessex) defeated the Vikings (Guthrum) it was here that Alfred accepted the surrender of the Vikings. It was at Wedmore that it was agreed, as a condition of the surrender, that the Vikings would leave this land and Guthrum would accept Christianity and be baptised. This little village becomes so much more important when you know that the whole course of English history changed here, it feels different, your walk here becomes so much more rewarding.

Adventure

Every little bit of learning rewards your efforts massively. Walk, cycle or paraglide to a place, know something of its history when you get there and walk, cycle or paraglide home. You don't have to be an expert; you do not have to have read War and Peace.

Somerset, which is a special place, is somehow enriched by the knowledge that Samuel Taylor Coleridge was inspired by the place, you will enjoy Somerset and the Southwest more for having read his Poems or his Poems more for having walked on The Quantock hills. Either way it only takes a little bit of learning to bring a place to life, to stir an emotion. Every part of the world has something more to offer, I cannot just take it on face value, I need a little more.

I spent years canoeing on the Lac de Leon and its tributaries in Southwest France, but the place literally came to life for me when I read about the Courant D'Huchet (the Huchet Stream) which links this lake to the sea. It was 'discovered by the Italian Poet Gabrielle Annunzio in 1908 and is often described by the locals as the 'Little Amazon'. When you canoe on this 'Little Amazon' you can suspend disbelief and let your imagination transport you away.

Adventure

The Courant D'Huchet 2011

You see things and things see you, Coypu, Lac de Leon 2011

When you feel a bit more confident

The key to the bigger adventures is the planning, it is quite possible that you will find this all a bit of a chore. Those of you, like me, who enjoy the planning, then the big adventures are there for the taking.

You may even have 'geared' some of your daily routine to the planning process. I always have a roof-rack on the car, I always have straps in the boot, I have lots of OS maps. Some of you will understand this straight away, others won't. I am lucky I have space at home, my canoeing stuff is organised and ready to grab as is my Diving gear and my Surfing Gear, my walking boots are always clean and waterproofed. All of this means that it is easy to plan. My friend Richard just contacted me to say 'how about a one-day canoeing trip on the River Wye on the 21st June? The day is booked, the transportation is sorted, and I know which three bags to throw in the boot of the car. Obviously,

Adventure

unlike Peter, I shall remember the Canoe.
Suffice to say that nothing aids the planning process like a bit of organisation.

For a good cause

At the age of 11 our son Wil came home from school one day, having seen a presentation about Autism. He had decided that he wanted to do something to raise money for the National Autism Charity. But what?

We sat and discussed what he could do that would be 'fun' and bit of a challenge, memorable and which could raise money. Eventually he hit upon the idea of cycling 100 miles of The Tarka Trail in Devon. He planned his route, contacted camp sites, practiced packing his bike and cycling with it fully loaded and eventually the start day came.

We fitted his bicycle with a trip computer:

Day 1

33

Adventure

Of course, his packing had to include food, camping gear and much much more.

He planned to do 20 miles a day, rain or shine, and he had to endure both, but he planned for this.

Prepared for shine

Ready for rain

Some days were tougher than others and the weather was, at times, grim, this is the United Kingdom after all. But when he finished, he had completed a mammoth adventure, he had developed an idea, planned it, executed it perfectly and raised a lot of money for a good cause in the process. His pride when he had finished was enormous.

Adventure

Well-earned fatigue, Wil Male 2009 having completed 100 miles of The Tarka Trail to raise money for the National Autism Society

Proof

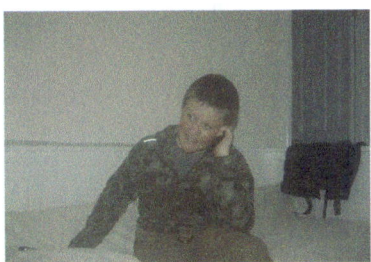

Telling his mum that he had succeeded. Every adventure you have can bring this feeling.

Adventure

These adventures and these successes are easy to ignore but the pleasure we can gain from them is enormous. Whether it is a cycle ride or teaching yourself to bake bread, or going for a walk in a crowded place, a well-planned adventure is an opportunity to succeed and celebrate.

Toilet Paper

For any adventure of more than one day, then the planning process becomes critical, a City Break needs a hotel or accommodation, two-day hikes will need to be timed for accommodation or campsites not to mention food and toilets. I have mentioned toilets a few times, forget to plan for them (or toilet paper) at your peril.

Some of the longer adventures

You may be familiar with the motorcycle tour that Chris and I undertook of the four compass points of the UK (The Liquorice Road, Dominic Male, Amazon) or The 500km solo Canoe trip across France (Reflections on Water, Dominic Male, Amazon). These were both pretty big undertakings but there are others, not quite so onerous but each of which, equally rewarding.

Pamplona or bust

Friends since University days, Chris and I had, in 1989 hitch-hiked to Pamplona from London in order to attend the Fiesta de San Fermin (The Bull Run). This trip was not terribly well planned. We

Adventure

aborted Hitch-Hiking at Bordeaux, in favour of the train (where we found two Americans), we gained the use of a car by chance to journey over the Pyrenees, we lost the two Americans in Pamplona (always a silver lining), we found some brilliant Australians and we eventually ended up back in Paris for the 200th anniversary of Bastille day, drinking Champagne on the steps of Sacre Coeur, being shouted at by a German, and wondering where we were going to sleep.

Earlier on this trip, however, whilst standing by the side of the N137 in the village of Clisson we had remarked that one day we would do this journey by motorbike. It took us 21 years to get organised, but organised we got.

We dusted off our aging motorbikes and we planned. By the time we met in Portsmouth to catch the ferry we were brilliantly organised, the journey was fantastic. We had arranged to meet with old friends in Pamplona, we had even arranged safe storage for our bikes. The festival was awesome, and we had no stress of bike storage or accommodation and all of the planning meant that when things went wrong (which they did) we were prepared. Riding our bikes was a pleasure, each new experience was a welcome part of the adventure.

Our bikes were expertly packed, we had practiced every packing permutation, we had everything we needed and each time we had to re-pack the bikes it was easy and 'automatic', we explored new places,

Adventure

saw new scenery, covered well over 1000 miles (1600kms) in relative comfort

The legendary 750cc Africa Twin, oh and Chris 2010

Expertly Packed 1340cc FXDB 2010

Adventure

Canoeing trips, like motorcycle journeys are a great escape for me. One can often find calm and solitude on these journeys whilst also getting to spend time with friends.

A little Adrenaline

For the same reasons I enjoy, both Diving and Skiing.

With Diving one meets people who share a passion, this brings with it a great camaraderie. You can, I guarantee, walk into any Dive Centre in the world and find people who are keen to help you plan great adventures, adventures which, even for seasoned, advanced divers, will bring awe and amazement, the planning of which will only enhance the experience. I have enjoyed diving adventures on the wrecks of Russian Mine Sweepers off the coast of Cuba, Swam with Sharks and into Jungle Caves (Cenote) in Mexico, wondered at the colour and wildlife on the reefs in the Egyptian Red Sea as well as swimming with Giant Turtles and Whale Sharks between the Islands of Cape Verde.

Adventure

Cave (Cenote) Diving, Mayan Peninsula, Mexico 2017, here, if you don't plan, you die

Adventure

When Emma and I go walking

The best adventure ever was exploring and discovering the stunning Levada walks of Madeira with my lovely Emma. Walking high on the mountain sides of Madeira through the stunning plant filled landscape and savouring the incredible vistas this place has to offer.

Walking the Levada with Em 2016

Adventure

Walking the Levada with Em 2016

In the mountains with My daughter Natalie

With family and friends, we have Hiked, Skied and Snowboarded from 4000m peaks in Colorado, through forests in British Colombia and over mountain top valleys in the French Alps. Part of each of these great adventures was meticulous planning right down to charging the Walkie-Talkie batteries, carrying the Kendal Mint-Cake and can of oxygen. Adventures are great alone but spectacular when shared. Even if you don't speak, even if you just look at each other and knowingly smile.

Adventure

The Rockies 2018

My snow buddy, Natalie 2018

Adventure

The key is not to rush into anything, an impromptu trip, is not an adventure. Find the things you like to do and start small, no one adventure is more 'worthy' than another. If planning a trip to your favourite train station to 'spot trains' is what you like, then fantastic, do it, do it well and do it as much as you can. Likewise, if sliding down K2 on a drink tray is what makes you buzz, then buzz away to your heart's content. But plan it.

With my students we often talk about planning, I remind them that a plan is there to let you know when things 'are not going to plan' and a good plan will ensure that when 'things are not going to plan, it is not a disaster.

"It does not do to leave a live dragon out of your calculations, if you live near one." – J.R.R. Tolkien

Another great Motorcycling Adventure with my adventure buddy, Chris

The Catalan Pyrenees.

Whilst Chris and I have undertaken several Adventures together from Hitch-Hiking through France to Circumnavigating the UK Mainland. A few years ago, Chris brought to my attention, the fact that the rules relating to riding 'dirt bikes' is very different in Spain than it is in the UK. Most notably, you ARE permitted to ride your motorcycle on any trails or tacks, as long as they appear on a map.

Adventure

In addition to this there are organisations who can provide all sorts of information to help plan. One we did was a form of orienteering challenge where we could buy a book of clues and a corresponding map and plan our motorcycle riding from there. This was too good to be true. We both possessed little 250cc trail bikes, so we set about planning. This was multi-faceted, we needed to:

- Prepare the bikes
- Plan to transport the bikes, which meant sourcing a trailer
- Get some off-road training
- Get the Clues and Maps
- Practice loading the bikes
- Plan the routes
- Get there
- Ride around in the mountains for a few days
- Have an adventure

Well, this was a disaster, we did not take spare bikes, so guess what, things could not have gone worse, in fairness we got there but my bike lasted less than two hours of the first day. A seemingly unresolvable electrical fault brought my adventure to a very early end. The bike was put back on the trailer and that was that, fortunately we had planned for disaster, and we had a means by which to get the bike home, Chris managed to ride the second day with others, whilst I explored the mountains in my car and on foot. But the initial three hours of riding and a great deal of patience on Chris' part had whet the appetite and the following

Adventure

year, we re-planned, took more spares and learnt a bit more about motorbikes. New routes into the mountains were mapped and a few days skirting Andorra via the high mountain passes were set to be awesome, and they did not disappoint.

We loaded up the trailer and enjoyed a brilliant road-trip down through France, over the border and into the Catalan Pyrenees.

We actually got two bikes on here

Once in the mountains we rode for days on tracks, over mountain peaks and through rivers, we discovered villages lost in time. Stopped for bread and wine picnics, only to be joined by wild horses, we rode for days without seeing another soul.

Adventure

Chris and I stopping for a well-earned rest 2019

All the time, following our planned trails and intersecting with civilisation only, and I mean only, when planned.

Me looking for electrical faults, Catalan Pyrenees 2019

When we had finished riding, Chris and I looked at each other like two gleeful children and Chris yelled at the top of his lungs "WE DID IT". Best motorcycle adventure ever.

Adventure

Chris and Dom, somewhere up a Catalonian Mountain 2019

Just one day's ridding 2019

6

What does this all mean?

Some will choose to live life passively, to happily bimble along with what they have and be content. If this is you, this is fine. If, however, you want a little more or like me, you get a little bored, you can choose to approach everything you do as a marvellous adventure, you can seek out the new and relish every success. This, for me at least, makes life a little more exciting and a bit more fun.

But remember most TV adventure is theatre, you baking a Soufflé is probably more risky and more difficult. Every day of your life can be an adventure, it is entirely up to you.

Go on a holiday, travel halfway around the world and sit by the pool, this is a trip. Find out about the place, go and see some stuff and discover a place, this is an adventure.

Adventure

One of my heroes, Ray Mears 2017, a man who doesn't make a big 'Song and Dance' about things, he just gets on with it, well planned, well thought out and obviously loving every second.

L - #0055 - 250821 - C50 - 175/108/3 - PB - DID3150262